AI Tools for Military Readiness

PETER SCHIRMER, JASMIN LÉVEILLÉ

RAND NATIONAL DEFENSE RESEARCH INSTITUTE

For more information on this publication, visit **www.rand.org/t/RRA449-1**.

About RAND

The RAND Corporation is a research organization that develops solutions to public policy challenges to help make communities throughout the world safer and more secure, healthier and more prosperous. RAND is nonprofit, nonpartisan, and committed to the public interest. To learn more about RAND, visit www.rand.org.

Research Integrity

Our mission to help improve policy and decisionmaking through research and analysis is enabled through our core values of quality and objectivity and our unwavering commitment to the highest level of integrity and ethical behavior. To help ensure our research and analysis are rigorous, objective, and nonpartisan, we subject our research publications to a robust and exacting quality-assurance process; avoid both the appearance and reality of financial and other conflicts of interest through staff training, project screening, and a policy of mandatory disclosure; and pursue transparency in our research engagements through our commitment to the open publication of our research findings and recommendations, disclosure of the source of funding of published research, and policies to ensure intellectual independence. For more information, visit www.rand.org/about/principles.

RAND's publications do not necessarily reflect the opinions of its research clients and sponsors.

Published by the RAND Corporation, Santa Monica, Calif.
© 2021 RAND Corporation
RAND® is a registered trademark.

Library of Congress Cataloging-in-Publication Data is available for this publication.

ISBN: 978-1-9774-0749-8

Cover: Pentagon: DoraDalton/Getty Images/iStockphoto. Source code: tihomir_todorov/Getty Images/iStockphoto

About This Report

All branches of the U.S. military collect monthly information on the readiness of their combat units, headquarters, commands, and other organizations. Military units and organizations report each month an overall readiness level, as well as readiness in four areas—personnel, equipment on hand, operational capability of that equipment, and training. These readiness levels are quantified on a numerical scale. The quantified readiness assessments are also accompanied by narratives explaining what is occurring in the unit that is affecting current or future readiness, either positively or negatively. The goal of this report is to use the textual, unstructured narratives to calculate overall readiness. One important benefit of such a capability would be to enable senior leaders to estimate how the readiness level of a unit could be affected by personnel, equipment, or training factors; another benefit would be to eventually have some automated, real-time interaction with unit commanders as they write their narratives in order to refine the information they provide and better align the narratives with the reported readiness levels.

This report will be of interest to people who seek to use artificial intelligence and its enabling technologies to support the mission of the Department of Defense and the military services. The report assumes the audience has a foundational understanding of machine learning and natural language processing.

The research reported here was completed in May 2021 and underwent security review with the sponsor and the Defense Office of Prepublication and Security Review before public release.

This research was sponsored by the Office of the Secretary of Defense and conducted within the Forces and Resources Policy Center of the RAND Corporation's National Security Research Division (NSRD), which operates the RAND National Defense Research Institute (NDRI), a federally funded research and development center (FFRDC) sponsored by the Office of the Secretary of Defense, the Joint Staff, the Unified Combatant Commands, the Navy, the Marine Corps, the defense agencies, and the defense intelligence enterprise.

For more information on the Forces and Resources Policy Center, see www.rand.org/nsrd/frp or contact the director (contact information is provided on the webpage).

Contents

Figures and Tables

Figures

Tables

Summary

Readiness is a perennial priority of America's military and civilian leaders, and is a cornerstone of our national security. Key to managing and improving readiness is being able to measure it. This gives our leaders not only situational awareness but also tools for exploring trade-offs with other priorities, such as modernization, force structure, and responsible use of the nation's resources. There are likely many ways in which artificial intelligence (AI) can eventually improve measurement and management of military readiness. The work discussed in this report advances the capability of computers to "understand" human language, describing factors that promote or impede readiness. Such a capability will facilitate the human-machine interface and could lead to more accurate measures of readiness and possibly even improve the management of readiness—although that is a more long-term goal.

Military units and organizations are required to submit monthly reports with an assessment of their overall readiness to fulfill their core missions and perform their designed capabilities. The overall readiness level of the unit is called its C-level, and is assessed on a scale from 1 (highest) to 4 (lowest).[1] In addition to these categorical assessments, unit commanders are also required to provide narrative comments about conditions and activities within their units that explain their reasoning behind the reported C-level. These comments address topics such as unit personnel, equipment availability and operational

---

[1] Occasionally a unit can be rated at C-level 5, but that is under rare circumstances, such as when a ground unit is converting from light infantry to armor and is swapping out equipment and personnel.

status, and training activities. The data set provides a wealth of information relating military readiness to a range of factors, described in the military's version of plain English.

Our research team built models, based on a deep neural network architecture, that predict the readiness level of units and organizations of one military branch relying solely on a free-text description of the current conditions and activities in a unit. Models were evaluated in terms of recall, precision, and F1 score, which are standard metrics to measure classification accuracy. *Recall* measures the fraction of actual targets (true positives plus false negatives) that are true positives. *Precision* measures the fraction of predicted targets (true positives plus false positives) that are true positives. *F1* measures the harmonic mean of the two. The overall accuracy of the best-performing model, measured in terms of its overall F1 score, is 75 percent. The accuracy metrics for each of the four C-levels are listed in Table S.1 and indicate high accuracy in all cases. The best model even did well at differentiating between the middle categories of the scale, which can prove difficult on text classification tasks.

In the course of building these predictive models, the team also built a language model for defense and national security, trained on over 2,500 RAND Corporation federally funded research and development center publications. The defense-specific language model (more precisely, word embeddings) was developed using the word2vec architecture and performs well on defense-specific word analogy and word relatedness tasks, outperforming pretrained general language word embeddings available in the public domain. Public domain word embeddings, trained on corpora such as Google News or Wikipedia,

Table S.1
Final Model Performance Metrics on Each C-Level

C-Level	F1 (Percentage)	Recall (Percentage)	Precision (Percentage)
C-1	73	70	77
C-2	76	71	74
C-3	71	69	73
C-4	74	72	77

are good general-purpose word embeddings useful on many natural language processing tasks, but as Tables S.2 and S.3 show, such embeddings do not work as well in a domain such as national defense that has a specialized vocabulary and semantic context. Table S.2 compares the results of a word relatedness task for the public domain and defense-specific embeddings, and Table S.3 compares the results of a word analogy task. Both tables list the top responses generated by the embeddings for the task at hand. For the analogy task, if we judged a term to be the correct answer, we indicate that in boldface.

Models capable of "understanding" the language of military readiness and national security could be incorporated into a range of tools, from a chatbot within a readiness reporting system to question answering and search tools. As a start, it could be integrated into the service branch's readiness reporting system on an experimental basis to gently prod unit commanders to provide more accurate and detailed comments. In the case of service readiness, the AI assistant could detect

Table S.2
Comparison of Word Embeddings on Task-Specific Word Relatedness

Comparison Word	Public Domain	Defense-Specific
readiness	capability	unit_readiness
	willingness	effectiveness
	commitment	operational
retrograde	anterograde	Class_IX
	prograde	depot_repair
	anomalous	redeployment
equipment	materials	materiel
	systems	ammunition
	hardware	supplies
training	preparation	collective_training
	skills	instruction
	teaching	on-the-job_training
tactics	tactic	TTPs
	strategy	techniques
	methods	operational_concepts

NOTE: TTPs = tactics, techniques, and procedures.

Table S.3
Comparison of Word Embeddings on Task-Specific Word Analogies

Analogy	Public Domain	Defense-Specific
VADM:Navy::LTG: _____	naval	**Air_Force**
	corps	**Army**
	command	**Marine_Corps**
Knox:Army::Randolph: _____	corps	**Air_Force**
	command	Armys
	forces	Marine_Corps
DDG:destroyer::FFG: _____	destroyers	**frigate**
	pns	Virginia_class
	guided-missile	surface_combatant
F/A-18:Navy::F-22: _____	N/A	**Air_Force**
		DoD
		MOD
Class_V::ammunition: Class_IX: _____	N/A	**spare_parts**
		materiel
		munitions

NOTES: Correct answers are indicted in boldface. ADM = admiral; DDG = guided missile destroyer; DoD = Department of Defense; FFG = frigate; LTG = lieutenant general; MOD = Ministry of Defense; VADM = vice admiral.

a mismatch between the commander's comments and the C-level by comparing the latter with the C-level predicted by the model. It could then ask for greater elaboration to justify the reported C-level. Over the longer term, an AI system capable of understanding the language of readiness might provide a human-interpretable summarization of readiness trends and inputs or enable sophisticated queries of commanders' comments. Such tools, capable of what is known as natural language processing or, more recently, natural language understanding, are an important branch of AI. While such tools are widespread in other domains, RAND has an opportunity to help advance the state of the art in the national security domain, which could lead to more accurate measures of military readiness and better information about unit-level factors that promote or impede readiness.

Acknowledgments

We are grateful to several RAND Corporation colleagues who contributed to this report. Gary Briggs, Karl Hutchinson, and Mitch Tuller provided essential assistance in getting the necessary software installed to enable this work. Benson Wong gathered the machine-readable federally funded research and development center (FFRDC) publications that were the basis for the defense-specific word embeddings we developed. Michael Linick and Rand Waltzman provided very helpful reviews of this document; Rand also provided interim feedback while the work was underway.

Abbreviations

1D	one-dimensional
ADM	admiral
CBOW	continuous bag of words
CNN	convolutional neural network
CPEST	collective, personnel, equipment readiness, supply, training proficiency
CPSRT	collective, personnel, supply, equipment readiness, training proficiency
DDG	guided missile destroyer
DoD	Department of Defense
DRRS	Defense Readiness Reporting System
DS	defense-specific
FFG	Frigate
FFRDC	federally funded research and development center
GRU	gated recurrent unit
HQDA	Headquarters, Department of the Army
IR	information retrieval
LR	logistic regression

LSTM	long short-term memory
LTG	lieutenant general
METL	Mission Essential Task List
MOD	Ministry of Defense
NLP	natural language processing
OH	one-hot
RNN	recurrent neural network
TF-IDF	term frequency–inverse document frequency
TTPs	tactics, techniques, and procedures
VADM	vice admiral

CHAPTER ONE

Interpreting Military Unit Readiness Through Machine Learning

Military readiness is a timeless priority for the United States. It reflects the ability of not just combat units but any type of unit or organization to perform its essential tasks and missions in defense of the nation. Civilian and military leaders want to ensure that they appropriately resource all of the factors that build and sustain readiness, and they strive to understand how these factors affect readiness within individual units and across the force.

How the Military Reports Readiness

The Defense Readiness Reporting System (DRRS) is the Department of Defense's (DoD's) official readiness reporting database and contains monthly readiness reports for each unit, organization, and installation in the Air Force, Army, Coast Guard, Marine Corps, Navy, and Space Force, as well as joint organizations, combat support agencies, and Joint National Guard units. Commanders are required to submit a report each month with an "assessment of their unit's ability to undertake their wartime missions for which they are organized or designed."[1] The overall readiness level of the unit is its collective assessment level (abbreviated as C-level) because it takes into account several contributing factors. The four factors, or "measured areas," that support overall readiness are personnel (P), equipment and supplies on hand (S),

---

[1] Joint Staff, *Force Readiness Reporting*, Washington, D.C.: Joint Staff, CJCSI 3401.02B, July 17, 2014. As of the publication of this report, the instruction had not been updated to include the Space Force but presumably will be.

equipment readiness and serviceability (R for the Air Force, Army, and Marine Corps; E for the Navy), and unit training proficiency (T). Collectively, they are abbreviated CPEST for the Navy and CPSRT for the other services. Like the overall C-level, each of these measured areas (P-level; S-level; R-level or E-level; and T-level) is assessed on a scale from 1 (highest) to 4 (lowest).[2] In addition to these categorical assessments, unit commanders are also required to provide comments about the factors affecting readiness within their units. Although the monthly readiness reports include more than just the CPEST/CPSRT assessments and commander comments, these are the focus of this modeling effort.

The Joint Staff and the services strive to establish objective criteria for assessing the level for each of the measured areas. For example, the Joint Staff–directed metrics for determining a P-level within a unit are reproduced in Table 1.1. The method for determining the S-level is similarly formulaic; both are essentially objective counting exercises. In theory, the R-level is, too; it is based on the number of items that are fully mission capable in accord with their technical manuals and includes separate calculations for different types of equipment, including a special category of major end items (such as tanks in an armor unit or airframes in an aviation unit). The T-level is easily the most subjective of the four supporting factors. Joint Staff and service instructions emphasize "designated conditions," "measurable standards," and "training events that must be completed at specified intervals," but it is an ongoing challenge to objectively measure unit training readiness.[3] Tasks vary from unit to unit; the conditions and terrain where units train vary by installation; and there is a constant cycling of personnel into and out of units. But the biggest challenge is the fact that how well a group of people performs a collection of complicated activities does

---

[2]   Technically, there are five possible levels that can be reported for the C-level and six possible levels for PEST/PSRT, but these are meant to be exceptional. A rating of 5 indicates the unit is undergoing some change in resourcing or equipping or is part of a directed program, such as when an Army or Marine Corps battalion is converting from infantry to armor; a rating of 6 indicates that the area is not measurable.

[3]   Joint Staff, 2014.

Table 1.1
Metrics for Determining the Personnel Level

Level	Available Strength (Percentage)	Critical Personnel MOS or Specialty (Percentage)	Critical Grade Fill (Optional) (Percentage)
1	100–90	100–85	100–85
2	89–80	84–75	84–75
3	79–70	74–65	74–65
4	69 or less	64 or less	64 or less

SOURCE: Joint Staff, 2014, Table 1.

not lend itself to binary, yes-or-no measures. There is some irreducible subjectivity required to make the assessment.

Unit commanders may not alter the PEST/PSRT levels prescribed by their respective formulas. The overall C-level is initially set at the lowest of the four resource areas, but commanders may subjectively upgrade or downgrade this initial assessment to best align the reported C-level with the definitions listed in Table 1.2. Tables 1.1 and 1.2 provide an interesting contrast and reflect a significant challenge in understanding military readiness. While the P-level—and, indeed, all the measured areas—have precise thresholds, and service-specific guidance even goes so far as to require special look-up tables to calculate certain PEST/PSRT levels, the C-level leaves plenty of room for interpretation. What does it mean to undertake "most of the mission," and how does that differ from undertaking "many, but not all, portions of the mission"? The challenges of assessing training readiness, which are literally part of C-level definitions, have already been addressed.

For the units examined in this study, special approvals are required to adjust the C-level by more than one level. When the commanders do make an upgrade or downgrade, they presumably explain this in their monthly comments. If they do not, the categorical readiness level will appear incongruous with the written comments. In fact, even if the commander is not subjectively altering the C-level, he or she may not clearly explain how resources and other factors affect unit readiness.

Table 1.2
C-Level Definitions

C-Level	Definition
C-1	The unit possesses the required resources and is trained to undertake the full wartime missions for which it is organized or designed.
C-2	The unit possesses the required resources and is trained to undertake most of the wartime missions for which it is organized or designed.
C-3	The unit possesses the required resources and is trained to undertake many, but not all, portions of the wartime missions for which it is organized or designed.
C-4	The unit requires additional resources or training to undertake its wartime missions, but may be directed to undertake portions of its wartime missions with resources on hand.

SOURCE: Joint Staff, 2014, Enclosure C.

The commander's comments are so important that there is even a section in an Army regulation titled "Importance of Commander Comments." The regulation states that the comments "are closely reviewed routinely by resource managers and senior leaders at higher headquarters, to include HQDA [Headquarters, Department of the Army], to identify urgent concerns requiring immediate actions."[4] The Air Force requires that each Major Command has a Readiness Office responsible for ensuring readiness reporting requirements, with responsibilities that include sampling, reviewing, and assessing the "adequacy of unit remarks." Those remarks deemed inadequate are to be challenged and corrected.[5] Marine Corps instructions state that "commanders should provide amplifying remarks . . . to assist with resourcing, training, and force management decisions."[6] Reading and interpreting these comments becomes a problem at scale because each service branch has

---

[4] Department of the Army, *Army Unit Status Reporting and Force Registration*, Washington, D.C.: Department of the Army, AR 220-1, April 15, 2010.

[5] Department of the Air Force, *Force Readiness Reporting*, Washington, D.C.: Department of the Air Force, AFI 10-201, March 3, 2016.

[6] Department of the Navy, *Marine Corps Readiness Reporting*, Washington, D.C.: Department of the Navy, MCO 3000.13A, July 18, 2017.

many hundreds or even thousands of units and organizations reporting readiness each month, and the comments can sometimes be extensive. With the massive task of reviewing readiness comments, time is a factor, but so is fatigue, and it could be extremely challenging to make connections and infer patterns from readiness reports over time or across different types of units. In short, what humans alone can do with the commander comments on the scale of a military service, let alone all of DoD, is limited.

How Machine Learning Tools Can Help

Computer automation is invaluable for tasks where requirements for scale, speed, and/or accuracy exceed human capabilities. While this has been generally true for decades, it has only been quite recently that computers have been able to perform at acceptable levels on challenging natural language processing (NLP) tasks. In prior work, RAND researchers used classical NLP techniques to build a model capable of interpreting the commanders' comments for Navy guided missile destroyers (DDGs) and estimating the ships' readiness levels. The research discussed in this report similarly attempts to predict unit readiness based on commanders' comments for one military service branch (henceforth "the service"). The current effort is more challenging in two ways, and brings to bear more advanced machine learning tools. It is more challenging because we are building analytic tools for all types of units and organizations within the service as opposed to a single class of Navy ship. It is also more challenging because it has an additional readiness category to deal with. The Navy study was based on readiness reporting in DRRS-Strategic, which uses only a three-tier scale (*yes, qualified yes,* and *no*). As one might imagine, the model in the prior Navy work more successfully identified the *yes* and *no* cases; the *qualified yes* cases proved more difficult. The F1 scores (which are a sophisticated way of calculating model accuracy) were 87 percent for the *yes* cases and 79 percent for the *no* cases, but just 60 percent for the *qualified yes* cases. This makes sense, because a *qualified yes* is more ambiguous—it is somewhere between a *yes* and a *no*. The service study described in this report uses the multitier C-level reporting rubric employed in the Air Force Input Tool, DRRS-Army, DRRS-Marine

Corps, and DRRS-Navy. This scale in a sense splits the middle ground into two readiness categories (C-levels 2 and 3). However, we also bring new tools to the effort. The prior study did not encode the commanders' comments using word vectors, and the model was a single-layer network logistic regression (LR).[7] This follow-on study uses word embeddings and deep neural networks. As a result, the best models for the service have an F1 score of about 75 percent, which is on par with the earlier Navy effort and far exceeds pure guessing based on the most frequent C-level.

Narrowly defined, the objective of this research is to train deep neural networks to calculate what a unit's C-level is based solely on the commander's comments. If the tool were only used to identify the correct C-level when the reported C-level is already available, it would be little more than a modern-day parlor trick. But viewed more broadly, this is a tool for interpreting the readiness effects of personnel, equipment, and training factors as described in service-specific language. Through the process of transfer learning, such a tool could be integrated into applications with AI-like capabilities. For example, a predictive model could be integrated into a chatbot that interacts with unit commanders through the computer interface with which readiness reports are submitted. Those commanders who are not are paragons of exposition and clarity might get feedback like, "Based on your comments, there is a 65 percent chance your unit is C-3, but you report C-2. Can you elaborate on why your unit is C-2?" A very simple chatbot such as this could be built using not much more than the model we have already estimated. Something more sophisticated would include a model to predict future readiness, a real-time language-generation model, and a way of parsing comments for specific details worth exploring, such as how a unit is preparing for a rotation at a training center or dealing with certain equipment shortages.

But the model need not be associated with DRRS at all. Although it uses the readiness reporting data for learning, the model is best con-

---

[7] Other models, such as Naive Bayes classifiers, support vector machines, and random forests, were also explored, and all produced about the same level of accuracy. The LR model was selected for its interpretability.

sidered as a tool for interpreting unstructured information about things like personnel turnover, equipment shortages, and training activities and deducing their likely impact on unit readiness. One potential next step might be to leverage this predictive model to develop tools that help service leaders understand how various factors affect unit readiness and how those factors should be resourced. In fact, that is one of the main goals of having commanders provide monthly comments about their units' readiness. The architecture developed in this project may be considered an important step in that direction, and if (like the chatbot example) it is integrated with other analytic tools, it can be an important element in developing AI to support military readiness.[8]

The Structure of This Report

The model implemented in this project uses a word embedding method to associate a vector representation to each word in a commander's comment and maps the collection of vectors in a comment to C-levels using a deep neural network. Chapters Two and Three describe the design and experiments with the word embedding method and deep neural network, respectively. Chapter Four concludes with recommendations for next steps.

---

[8] Since starting this work, new models have been introduced that significantly advance the state of the art across various NLP tasks—models such as BERT and, most recently, GPT-3. In light of this, the work described in this report can be seen as offering but a preview of the kinds of capabilities afforded by emerging NLP techniques.

CHAPTER TWO

Defense-Specific Word Embeddings

As a preliminary step to training deep neural networks to interpret unit readiness descriptions, we developed defense-specific word embeddings using word2vec, a method introduced in 2013 by Tomas Mikolov and colleagues.[1] Word2vec is a shallow network employing one of two approaches. The continuous bag of words (CBOW) approach takes a snapshot of text that is n words before and after a "target" word and attempts to predict the target word using unique weights for each of the surrounding words. Through gradient descent (backpropagation) on an objective function that captures the accuracy of the prediction, the weights for each word are gradually adjusted to better predict target words. The skip-gram approach similarly estimates word vectors but with an objective function that represents the opposite task: trying to predict n words before and after the target word. With either approach, when the model is finished training, the weights are preserved as vectors, each uniquely representing a different word in the vocabulary. Words that have similar meanings occur in similar contexts (i.e., have similar words in their n-word window) and therefore should have similar weights that are learned. The word vectors can then be used for downstream language processing tasks such as search, question answering, and text classification.

The Google team that developed the word2vec method for building word vectors published pretrained vectors trained on a portion of the Google News data set. The training data comprised roughly

---

[1] Tomas Mikolov, Kai Chen, Greg Corrado, and Jeffrey Dean, *Efficient Estimation of Word Representations in Vector Space*, arXiv, last revised September 7, 2013.

100 billion words.[2] In addition, other research teams published their own versions of pretrained vectors that might have had different training sets, or somewhat different model architectures, or both. These vectors are available for download and can be used for machine learning projects. But word vectors trained on general-purpose corpora may not capture the specific meanings of words from a specialized lexicon such as that of the military. Another issue with general-purpose public domain embeddings is that specialized words may simply not appear in general-purpose corpora, such that these words have no defined embedding. Some techniques help mitigate the impact of a few out-of-vocabulary words,[3] but the issue remains acute on very specialized text.

Because the language of military readiness is so domain-specific, we built custom word embeddings using the word2vec approach and then used those embeddings in the deep neural networks. Despite the specificity of the language used to generate the word embeddings, a drawback to this approach is the relative paucity of data on which to train. Compared to the size of the Google News corpus, the service readiness comment corpus is minuscule, so we boosted our training corpus with unclassified machine-readable documents published by RAND's national security and defense FFRDCs: the Arroyo Center, the National Security Research Division, and Project AIR FORCE. Our rationale is that the language used in these documents is much more specific to readiness and defense concepts and will likely not suffer from the same conflation of uses that the pretrained public domain vectors would. The combined text from these publications is more than ten times greater than the service readiness data, so our training data increased by more than an order of magnitude.

The FFRDC publications served a second purpose. Besides using them to boost the training data for service-readiness word embeddings, we also developed a set of defense-specific word embeddings just based on the FFRDC publications alone. These can be used for downstream

---

[2]   Google Code, "Word2vec," webpage, undated.

[3]   Mikhail Khodak, Nikunj Saunshi, Yingyu Liang, Tengyu Ma, Brandon Stewart, and Sanjeev Arora, "A la Carte Embedding: Cheap but Effective Induction of Semantic Feature Vectors," in *Proceedings of the 56th Annual Meeting of the Association for Computational Linguistics*, Stroudsburg, Penn.: Association for Computational Linguistics, 2018.

NLP tasks in the same way that pretrained public domain embeddings are. The following sections show how these embeddings, which we will refer to as defense-specific embeddings, compare with the public domain embeddings on various tasks.

Data

We gathered nearly 2,600 unclassified, machine-readable RAND FFRDC publications, including documents marked for official use only and terminal drafts, dating back to the 1990s. Some of the pre-processing steps performed on the data include stitching hyphenated words together, cleaning up odd characters, splitting up single tokens like *faith/hope/love* into separate tokens, and so forth. The documents were then split into sentences (groups of tokens) and randomly sorted so that sentences from various documents and various document types were mixed together. This was done because the method we used employs a declining learning rate throughout and across model training epochs,[4] and we did not want the order of the documents and document types to affect estimation of the model weights.

Method

As discussed above, there are two types of architectures for word2vec: CBOW and skip-gram. According to the Google research team, skip-gram works better on infrequent words but is slower. The options for the training algorithm are hierarchical softmax or negative sampling—both of which are approximations of a true softmax estimator that would predict the likelihood of every word in the vocabulary. Hierarchical softmax works better on infrequent words, but negative sampling is better with low-dimensional vectors.[5] Other options include the learning rate and the size of the sampling window.

There are also several options for processing the text to build the vectors. Text can be cased or uncased, and tokens may be only

---

[4] *Epochs* is a term of art, meaning passes over the data; models are trained by examining the data multiple times and updating their parameters on each pass.

[5] For a discussion of the performance of the various architectures and algorithms, see Google Code, undated.

unigrams (i.e., single words) or *n*-grams, combinations of *n* (typically no more than two or three) words. Even the tokenizer itself can treat things like hyphenated words or contractions differently. Finally, vectors can be set at any size desired, with typical sizes being 50, 100, and 300 elements. All embeddings assessed in this report are 100-element embeddings. Table 2.1 summarizes the model and data combination options, but there are also hyperparameters that can vary as well.

Each implementation of the word2vec model entails selecting one option from each column (for a total of 32 possible configurations), such as CBOW with negative sampling on uncased unigrams from the Treebank tokenizer, or skip-gram with hierarchical softmax on based bigrams from the Gensim tokenizer. The investigations reported here focus on comparing the performance of a subset of all possible configurations. Some of the best-performing configurations are listed in Table 2.2. A drawback with the cased and bigram options for the data is that they generate much larger files. With cased data, *Navy* and *navy* in the original text are considered two different tokens, each with its own vector. With bigrams, *united* and *states* are separate tokens, but so is *united_states*. Combining bigrams and cased tokens could create as many as five tokens: *united, states, United, States,* and *United_States*. Additionally, the model that built the bigrams must also be transferred to the target text if the bigrams are to be replicated in a transfer learning scenario; otherwise, having bigrams in the word embeddings is useless.

Results

We tested the defense-specific word embeddings on two intermediate tasks that explore the embeddings' ability to make semantic distinctions

Table 2.1
Model and Data Combinations for Defense-Specific Embeddings

Model Variants		Data Variants		
Architecture	Output Vector	Casing	Token Length	Tokenizer
CBOW	Negative sampling	Cased	Unigrams	Gensim
Skip-gram	Hierarchical softmax	Uncased	Bigrams	Treebank

between words and phrases, some of which are drawn from everyday language and some of which are drawn from defense-specific language. One of the tasks is an analogy task, and the other is a word recall task. We refer to these as intermediate tasks because they do not reflect the ultimate purpose of developing the embeddings. Rather, such tests provide practitioners a preview of how well embeddings might perform in final tasks, such as training an algorithm to understand the language of military readiness.[6] We were specifically interested in how well various implementations of our defense-specific word embeddings perform relative to one another and to a general-purpose public domain embedding. The defense-specific word embeddings examined in this chapter are based solely on the FFRDC publications and do not include text from the commanders' comments. Those were subsequently added in a classified computing environment. The public domain word embedding was developed by the Stanford University Natural Language Processing Group using Wikipedia and newswire data as training data and GloVe—which is closely related to the word2vec algorithm—as the training algorithm.[7] We compare the embeddings' performance on word analogy tasks and word relatedness tasks.[8]

The word analogy task assesses how well embeddings correctly calculate which word best completes an analogy such as Paris:France::Rome: _____. Researchers at Google curated a list of thousands of analogies for this task.[9] The analogies are grouped into several categories: world capitals, nationality, U.S. cities, family and gender, currency, adjective-adverb, comparatives, superlatives, present

---

[6]   An excellent example of how this is done can be found in Yada Pruksachatkun, Jason Phang, Haokun Liu, Phu Mon Htut, Xiaoyi Zhang, Richard Yuanzhe Pang, Clara Vania, Katharina Kann, and Samuel R. Bowman, "Intermediate-Task Transfer Learning with Pretrained Models for Natural Language Understanding: When and Why Does It Work?" in *Proceedings of the 58th Annual Meeting of the Association for Computational Linguistics*, Stroudsburg, Penn.: Association for Computational Linguistics, 2020.

[7]   A GloVe implementation is available at Jeffrey Pennington, Richard Socher, and Christopher D. Manning, "GloVe: Global Vectors for Word Representation," webpage, Stanford Natural Language Processing Group, August 2014b.

[8]   All of the models compared in this section have a dimensionality of 100.

[9]   The test set was originally described in Mikolov et al., 2013.

participles, past tense, plurals, and verb conjugation. One issue with using this test data set is that many of the analogies are arcane or irrelevant to our task. There may be some instances where it is useful to know that the correct answer to Kazakhstan:Astana::Burundi:_____ is Bujumbura, or that Tasty:Tastiest::Warm:_____ is Warmest, but, if so, one is better off using the public domain embeddings. Unsurprisingly, our defense-specific embeddings performed very poorly on this task, with only about 25-percent accuracy, varying slightly depending on the tokenization scheme. By comparison, the public domain embeddings achieved 65-percent accuracy. The gap closed slightly when we eliminated the most irrelevant categories such as world capitals, U.S. cities, and family.

Another quantifiable means of assessing embeddings is how well they perform on word relatedness tasks. The Israel Institute of Technology has built a list of 353 pairings of English words that humans assessed on a scale from 0 to 10 for their relatedness.[10] This is more subjective than the analogy task because the analogies are subject to rules of grammar and geography: the word for a female monarch is *queen*, not *king*, and Chicago is in Illinois, not Mississippi. With the word pairings, people may reasonably disagree over the relatedness of *journey* and *car* (average relatedness score 5.85) or of *planet* and *moon* (average relatedness score 8.08). The defense-specific embeddings had a Spearman correlation of roughly 0.45 with the word relatedness scores versus a 0.53 Spearman correlation for the public domain embeddings. Neither did particularly well, although as the examples above show, many of these terms, like the analogies, are not relevant to our task.

It is perhaps unsurprising that the defense-specific word embeddings did not perform particularly well—in an absolute sense, or even relative to the general-purpose public domain embeddings—on the intermediate tasks discussed thus far: the defense-specific embeddings

---

[10] The test set was originally described in Lev Finkelstein, Evgeniy Gabrilovich, Yossi Matias, Ehud Rivlin, Zach Solan, Gadi Wolfman, and Eytan Ruppin, "Placing Search in Context: The Concept Revisited," *ACM Transactions on Information Systems*, Vol. 20, No. 1, 2002, and is one of the data sets included in Gensim, which is a Python library for training vector embeddings and performing other NLP tasks. See "Gensim: Topic Modelling for Humans," webpage, undated.

were trained on a much smaller and narrower-focused corpus. This does not necessarily mean they would underperform on the ultimate task of training a model to interpret readiness comments. To understand how they might do on that task, and which particular implementation of the embeddings might work best, we identified a small set of specific defense-relevant terms and analogies and tested the embeddings on those. This approach is reminiscent of earlier work in which evaluation is conducted on a small set of words with particular relevance to a task of interest.[11] We did not have the time or resources to generate hundreds of word pairs and score them, or to think up thousands of analogies, but we came up with a reasonable list of items to put the embeddings through their paces. The determination about which embeddings perform best is based on subject matter knowledge, not on a quantifiable score, as with the analogies and word relatedness assessments. We found the most similar terms in each embedding for the following: *readiness, retrograde, training, personnel, equipment, mission, installation, qualified, tank, ammunition, maneuver, intelligence, tactics, ttps, METL, mission-essential, operations, defense, assault,* and *contact.* A partial list of results is given in Table 2.2. In some cases the public domain embeddings hold their own, but in general the defense-specific embeddings are clearly superior. A good example is how the defense-specific embeddings recognize the similarity between tactics and TTPs (tactics, techniques, and procedures). Curiously, the skip-gram hierarchical softmax model performed very poorly, even compared to the public domain model. This may be due to the relatively small size of the data set and the combination of architecture and output vector that are both sensitive to rare words.

We also created some analogies to test the embeddings, and the results are presented in Table 2.3, which (as was the case with Table 2.2) lists the top three answers for each analogy generated by the different embeddings. If one of the top answers is what we consider to be correct,

---

[11] William L. Hamilton, Jure Leskovec, and Dan Jurafsky, "Diachronic Word Embeddings Reveal Statistical Laws of Semantic Change," in *Proceedings of the 54th Annual Meeting of the Association for Computational Linguistics*, Stroudsburg, Penn.: Association for Computational Linguistics, 2016.

Table 2.2
Comparison of Word Embeddings on Task-Specific Word Relatedness

Comparison Word	Public Domain	CBOW Negative Sampling Unigram, Uncased	CBOW Negative Sampling Unigram, Cased	CBOW Negative Sampling Bigram, Cased	Skip-Gram Hierarchical Softmax Unigram, Uncased
readiness	capability	deployability	deployability	unit_readiness	requirements
	willingness	resiliency	effectiveness	effectiveness	c1/c2
	commitment	fitness	performance	operational	ar/arng
retrograde	anterograde	unserviceable	redeployment	Class_IX	receipting
	prograde	redeployment	shipment	depot_repair	unserviceable
	anomalous	shipment	unserviceable	redeployment	ssas
equipment	materials	materiel	materiel	materiel	dog-related
	systems	supplies	supplies	ammunition	refueling-specific
	hardware	hardware	ammunition	supplies	icpa
training	preparation	mentoring	instruction	collective_training	master-mechanic
	skills	trainers	mentoring	instruction	master-driver
	teaching	refresher	proficiency	on-the-job_training	bc3
tactics	tactic	ttps	TTPs	TTPs	bkthrgh
	strategy	techniques	techniques	techniques	6-11c/navy
	methods	doctrines	swarming	operational_concepts	3-2.64

NOTE: Uncased = lowercase words only; cased = upper- and lowercase words.

we indicate that answer in boldface. Interpreting the results may require some explanation. A vice admiral, abbreviated VADM, holds the same rank as a lieutenant general in the Air Force, Army (LTG is the Army's way of abbreviating it), and Marine Corps. Fort Knox is the location of the Army Human Resources Command, and Randolph Air Force Base is the location of the Air Force Personnel Center. A DDG is a Navy destroyer, and an FFG is a Navy frigate. The F/A-18 is a Navy Hornet combat aircraft, while the F-22 is an Air Force Raptor combat aircraft. The military has nine classes of supply, with ammunition designated as

Table 2.3
Comparison of Word Embeddings on Task-Specific Word Analogies

Analogy	Public Domain	CBOW Negative Sampling Bigram, Cased
VADM:Navy::LTG: _____	naval	**Air_Force**
	corps	**Army**
	command	**Marine_Corps**
Knox:Army::Randolph: _____	corps	**Air_Force**
	command	Armys
	forces	Marine_Corps
DDG:destroyer::FFG: _____	destroyers	**frigate**
	pns	Virginia_class
	guided-missile	surface_combatant
F/A-18:Navy::F-22: _____	N/A	**Air_Force**
		DoD
		MOD
Class_V::ammunition: Class_IX: _____	N/A	**spare_parts**
		materiel
		munitions

NOTE: Correct answers are indicted in boldface.

Class V and repair parts designated as Class IX. The defense-specific embeddings clearly outperformed the public domain embeddings on each of these analogies and, in fact, the public domain embeddings we used did not even include bigrams, so they could not complete the Class_V or F/A-18 analogies.

Discussion

On the intermediate, task-specific word relatedness and analogy tests we created, the defense-specific embeddings appear to significantly outperform the public domain embeddings and would likely be useful in downstream NLP models dealing with defense-related matters. In

Chapter Three we describe how we augmented the commanders' comments with the FFRDC texts to build word embeddings that were incorporated in the deep neural networks designed to interpret such comments.

Another use for the word embeddings developed here might be for query expansion as part of an information retrieval (IR) system. Query expansion refers to automatically generalizing a query so as to minimize chances of missing relevant information. A simple form of query expansion might be to retrieve not only a target keyword but also its synonyms, automatically obtained from a thesaurus. Word embeddings would instead enable query expansion based on word similarity (an approach that has been used in the medical literature),[12] such that the embeddings generated here would enable for military-relevant text data. The word embeddings could also help improve IR in other ways, including suggesting query completions and question answering.[13]

---

[12] Ferhat Aydın, Zehra Melce Hüsünbeyi, and Arzucan Özgür, "Automatic Query Generation Using Word Embeddings for Retrieving Passages Describing Experimental Methods," *Database*, Vol. 17, 2017.

[13] Kezban Dilek Onal, Ye Zhang, Ismail Sengor Altingovde, Md Mustafizur Rahman, Pinar Karagoz, Alex Braylan, Brandon Dang, Heng-Lu Chang, Henna Kim, Quinten McNamara, Aaron Angert, Edward Banner, Vivek Khetan, Tyler McDonnell, An Thanh Nguyen, Dan Xu, Byron C. Wallace, Maarten de Rijke, and Matthew Lease, "Neural Information Retrieval: At the End of the Early Years," *Information Retrieval Journal*, Vol. 21, 2018.

A Deep Neural Network to Estimate Readiness

This chapter introduces the deep neural network developed in this project to estimate C-levels of units and organizations in the service based on commanders' comments. The data were accessed in the business intelligence layer for the service's readiness reporting system and provide information about the readiness of all reporting units in the service's active and reserve components. The system is accessed through the Secure Internet Protocol Router Network, and the information within it is classified at the Secret level; it contains monthly readiness data. Besides the categorical ratings for CPEST/CPSRT and the commanders' comments, as discussed in Chapter One, the system contains information about units' mission essential task proficiency and other training data, the number of personnel assigned to the units, individual and subunit qualification, equipment on hand, and more.

Data

We gathered the categorical ratings for CPEST/CPSRT and commanders' comments for all active component units, organizations, and installations for fiscal years 2015–2019, inclusive, for a total of five years of data. There are rules for the types of units that are required to submit monthly readiness reports and under what circumstances. Even within a single military service, a wide variety of units and organizations submit reports: not only operational units but also military hospitals, personnel centers, analytic groups, supply depots, and military bases and installations.

Each observation in our data set was a unit month. For a full year, a single unit would generate 12 unit-month observations. Not all units report every month, and units are activated or deactivated over time, so there are not necessarily 60 observations for every unit in our five-year data set. In addition, units get reorganized from time to time, adding a new type of subordinate unit or swapping out some piece of equipment or even just moving to a new home station that might affect their training environment, so the relevant readiness information may change over time.

After we gathered the data, we randomized the order of the observations, kept 80 percent of the observations for modeling, and set aside 20 percent as an evaluation set. We did nothing with the evaluation set until we had completed all of our modeling. We split the modeling data into a training set and a validation set. The common practice is to train a model on a set of training data and see how well it performs on a set of validation data. What tends to happen with machine learning models is that they will begin to overfit their parameters to the training set over multiple training runs, or epochs, and their performance on the validation set will plateau or even peak and then decline in later epochs. Analysts performing hyperparameter fine-tuning or testing alternative model architectures focus on how the models perform on the validation set, not the training set. But overfitting can become a metaproblem in which the analyst selects the best combination of architecture and hyperparameters that fit the validation data without truly knowing how the model will perform on new data. That is the purpose of the evaluation data, and best practice dictates that it only be used once all modeling is complete, purely to evaluate how the model performs on new data. This is what we did.

After setting aside the evaluation set, we integrated the commanders' comments with the FFRDC corpus and recalculated the defense-specific word embeddings. The algorithm performs the CBOW and skip-gram predictions within—not across—individual sentences, so we randomly shuffled individual sentences from the commanders' comments with individual sentences from the FFRDC corpus, over-sampling commanders' comments because the total number of comments was less than the total number of sentences in the combined FFRDC corpus. The new defense-specific word embeddings were used in the embedding layer of the deep neural networks.

Model Metrics

In a categorization task, a truly naive model would simply predict the most frequent category in all cases. If there is relative balance among the classes, the model would perform extremely poorly, like guessing the outcome of a coin toss. Conversely, if there is great imbalance in the frequency of categories, the accuracy could be high, but the model would tell you nothing you did not already know. This is why raw accuracy—the number of correct predictions a model makes—is normally not a good measure of model performance for discrete categorization tasks. Unless otherwise specified, the performance metric discussed in this chapter is an F1 score calculated using two other measures, *recall* and *precision*. *Recall* measures the fraction of actual targets (true positives plus false negatives) that are true positives. *Precision* measures the fraction of predicted targets (true positives plus false positives) that are true positives. There is a trade-off between recall and precision. If a model is searching for needles in a haystack, it can achieve 100-percent recall simply by labeling everything a needle, because that would generate no false negatives. However, almost everything would be a false positive, and precision would be close to zero. Achieving 100-percent precision is not guaranteed, but the model could label the one thing it is most certain is a needle and call everything else hay. If the needle prediction is correct, there will be no false positives, but all the other needles would be false negatives, and recall would be close to zero. The F1 score calculates the harmonic mean of the two, and because a harmonic mean applies disproportionate weights to low values, a model can achieve a high F1 score only if recall and precision are both relatively high.

In addition to the F1 metric, we also present a confusion matrix for our modeling results. A confusion matrix is a table in which the rows contain actual categories (in this case, reported C-levels), and the columns represent the model predictions. The downward diagonal values indicate when the model correctly predicted the actual category, and the other values indicate where the model erred. The values in our confusion matrix are classified, because the rows would indicate the overall readiness level of service units, but we can show a heat map indicating not just relative accuracy of the model but also how it erred when it erred.

The Baseline Model

Our baseline model is an LR model and a shallow network in the sense that there is simply an input layer (the commanders' comments) and a single set of parameters that generate an output layer (the C-level probability); there are no hidden layers, as with a typical deep neural network. In the previous work on Navy readiness, the various shallow-network model options performed roughly same. We experimented with three methods of encoding the comments, described in detail in the appendix to this report. The first method was a simple one-hot (OH) encoding, the second method summed the defense-specific word vectors for each word in an observation, and the third method applied a term frequency–inverse document frequency (TF-IDF) weighting to the defense-specific word vectors in an observation.[1]

The LR model performed poorly regardless of how the comments were encoded. The OH encoding had an overall F1 score of 0.55, and the two defense-specific (DS) encoding variants performed even worse, with an overall F1 score of around 0.35. With four possible outcome categories, an F1 that low is not much better than random guessing. The LR-OH model—the best of the three—performed about equally poorly for all C-level categories; the individual F1 score for each category ranged from a low of 0.49 to a high of 0.61. In the prior Navy work, our LR-OH model had an overall F1 of around 0.74, but this was with only three categories, not four, and for a single type of ship. Clearly the service readiness task is much more challenging.

Deep Neural Network Models

Given this low baseline performance, we were confident that the deep neural networks would perform better, and we were pleased to see that they did. Our final model ended up with an overall F1 of 0.75, roughly the same as achieved on the much easier Navy task.

The software package we used was TensorFlow 1.13 on a quad-core central processing unit. TensorFlow has an abstraction library, Keras, on top of it that greatly simplifies model building. Our first

---

[1]   TF-IDF is a measure of how often a term is used relative to the number of documents in which it is used, and is broadly interpreted as representing a word's importance within a given document—or, in our case, a comment.

neural network was a very simple one using the defense-specific word embeddings, a global pooling layer, and a dense layer using a softmax estimator to predict the likelihood of each C-level for every observation. The likelihoods generated by the softmax estimator sum to 1.0, so an output of the model would be something like [0.11, 0.70, 0.12, 0.07] for C-levels 1 through 4. In this example, the categorical prediction would be C-level 2, since that is the highest probability. Normally the parameters of each layer are randomly initialized and then adjusted through backpropagation, but the Keras application programming interface enables assigning weights and then freezing one or more layers of a model. For this first model we assigned the defense-specific embeddings to the embedding layer and froze them to test their usefulness out of the box. This very basic model performed essentially the same as the baseline LR-DS variants, with an F1 score that peaked around 0.42 before the model began overfitting.[2]

If we initialize the embedding layer with the defense-specific embeddings but allow that layer to update, we get a jump in performance, with F1 climbing to about 0.62. With a vocabulary of around 200,000 words (more accurately, tokens) and a 100-element vector assigned to each token, the embedding layer has 20 million parameters. The global max pooling layer on top of it does not have any adjustable parameters but simply finds the maximum value for each vector element from all the words in the first layer, and the dense layer has 400 parameters (four possible categories multiplied by a 100-element vector from the pooling layer), so the simplest model only has 400 parameters to be adjusted. This is the same as with the LR-DS models in our baseline. Once we unfreeze the embedding layer, there are now 20 million more parameters that can be adjusted to improve model performance. With many orders of magnitude of additional parameters to adjust, it makes sense that this simple model would improve performance. As might be expected, it also takes much longer to run—a single epoch takes about nine minutes on the quad-core central processing unit instead of seven seconds. We also tried a variant in which we did not even bother initializing the embedding layer with defense-specific embeddings and instead allowed the

---

[2] As a reminder: unless otherwise specified, all F1 scores reported here were generated on the test set of the modeling data.

model to randomly initialize that layer and then update. This produced an even higher F1 score of about 0.71.

At this point, a pattern is already emerging: despite the impressive performance of the defense-specific embeddings on word relatedness and analogy tasks, they did not improve the performance of either the baseline LR model or the simple neural network model. On the contrary, these models actually did considerably worse when using the defense-specific embeddings. One hypothesis we had is that the defense-specific embeddings may help prevent the models from overfitting on the training data, but at best the overfitting was delayed by a few epochs when using defense-specific embeddings. As we explored additional model architectures, it became apparent that using randomized initial embedding weights and allowing them to update produced superior results. Nonetheless, the defense-specific embeddings may be useful for tasks other than text classification, and we will return to that idea in Chapter Four.

The problem with using a pooling layer directly on top of an embedding layer is that all word order is lost, and the inputs become just a glorified bag of words, in the sense that *the dog bit the man* is encoded the same as *the man bit the dog*, albeit in a high-dimension matrix. The way to retain word order is with recurrent neural networks (RNNs), and we next began stacking them on top of the embedding layer. More specifically, we used long short-term memory (LSTM) layers and their cousins, gated recurrent unit (GRU) layers. For short comments such as *unit is fully trained and ready*, a simple RNN may be sufficient, but the average comment is significantly longer, and simple RNNs do not perform as well. GRUs and LSTMs are variants of RNNs that are much better at processing longer sequences, and although GRUs have a simpler architecture than LSTMs they are not necessarily inferior when it comes to accuracy. They are a little faster, however.[3]

---

[3] Around the time this work began, new language models that significantly advanced the state of the art across various NLP tasks were developed. These language models are based on what is known as a transformer architecture and are much more computationally intensive than the models developed for this study. This can pose a considerable challenge for classified computing environments. Fortunately, less computationally intensive architectures have proven quite effective on text classification tasks, such as the one in this study.

One of the conceptual ideas behind deep neural networks is that one layer aggregates information from the layer below it into a bigger pattern. This kind of compositionality is probably better explained with image processing, though the same principle applies to NLP. The first layer of an image processing model may simply recognize darker and lighter shades, while the next layer recognizes lines and edges in the shading, the next layer assembles lines and edges into simple shapes, another layer assembles simple shapes into complex shapes like ears and whiskers and tails, and the final layer assembles ears and whiskers and tails into pictures of fully formed cats and dogs.

As of 2018, Google's online language translation tool was built atop a massive, seven-layer LSTM model.[4] We experimented with multiple layers of GRUs and LSTMs to see how performance improved, and found the improvements marginal at best. GRUs performed the same as LSTMs but were faster, so we experimented more with them by making them larger (in terms of their output parameters) and by stacking several atop one another. We also experimented with convolutional neural network (CNN) layers. The application of two-dimensional CNNs to image processing is well known, but CNNs have also proven useful for processing one-dimensional (1D) inputs like sequences. A 1D CNN in essence takes samples of a sequence of data, such as a commander's comments, and strings those samples together into a shorter sequence. Like RNNs, 1D CNNs can be stacked atop one another and produce increasingly shorter sequences. CNNs and GRUs/LSTMs can be interspersed as well.

Besides the model architectures, another potentially important aspect of model building is the optimizer. There are several available in the older version of TensorFlow we used, and several more are available in version 2.01, which is the current stable version as of this writing. We experimented and found that the Adam optimizer worked as well or better than RMSProp and stochastic gradient descent, two other popular optimizers. The default learning rate for Adam in TensorFlow is 1e-4, but we achieved slightly better performance with 1e-3.

---

[4] François Chollet, *Deep Learning with Python*, Shelter Island, N.Y.: Manning Publications, 2018.

We experimented with different batch sizes, the number of observations used by the model for one update of the weights via backpropagation. The optimal batch size varies by task. In this work, smaller batches yielded better estimates but took more time to complete a run. We found that model performance improved until batch sizes diminished to about 32, with smaller batches not yielding improvements worth the significant additional time to estimate the model.

In sum, there are innumerable combinations of architectures, optimizers, hyperparameters, and layer settings that go into model building, and we experimented with many of them. With the top-performing models, when we initialized the embedding layer with defense-specific word embeddings, the F1 score topped out in the range of 0.60–0.65; randomizing and updating the embeddings improved model performance from 5 to 10 percentage points pretty consistently. The best models approached an F1 of 0.75, which means a lot of work went into improving the score by 0.04 from the 0.71 that was achieved with the first, simple network with randomized embedding weights. Tweaking the learning rate of the optimizer squeezed out a slight improvement in F1, but the biggest difference that yielded a few percentage points' increase was adding the recurrent layer. The GRU worked as well as the LSTM and was faster, and adding additional recurrent layers made minimal difference. Our final model architecture is presented in Table 3.1.

Table 3.1
Final Model Architecture for Service Unit Readiness Prediction

Layer Type	Dropout	Dimensions	Number of Parameters
Embedding		(200,000, 100)	20,000,000
Conv1D		(32, 5)	16032
MaxPooling1D		32	
GRU	(0.2, 0.2)	128	87936
Dense		4	516

NOTES: Loss function: categorical cross-entropy; optimizer: Adam (learning rate = 0.001); batch size: 32.

Results

The F1 score for this final model is 0.751, the best of any we tested, though a number of architectures came close. This was the result for the test set of the model data. The final step is to see how it performs on the evaluation data that we initially set aside. The model performed nearly as well on the new data as it did on the model test set, with an F1 of 0.748. The overall recall of the model on the new data is 0.738, and the overall precision is 0.763. F1, recall, and precision scores can also be calculated for each C-level individually, and their calculations on the evaluation data are listed in Table 3.2, which shows that the model performs nearly equally well across all C-levels, even doing a very good job of differentiating between the two middle categories. Conversely, neither the highest nor the lowest readiness level is especially easy to identify.

Figure 3.1 is a heat map of the confusion matrix for the evaluation data. Because the underlying data are classified, we cannot show the actual number of units reporting at each C-level, but the heat map provides color coding of the highest and lowest numbers. Each cell value is normalized by the sum of its row and column (in essence, a matrix of F1 scores), and darker colors indicate a higher number. Because of the way confusion matrices are arrayed, with actual values in rows and predicted values in columns, high numbers along the diagonal from top left to bottom right are desired—and, indeed, that is what we see. The lightest areas of the heat map are at the top-right and bottom-left corners, meaning the model is least likely to predict a C-1 for a unit that is C-4, or vice versa, but this does occur for a small number of observations.

Table 3.2
Final Model Performance Metrics on Each C-Level

C-Level	F1 (Percentage)	Recall (Percentage)	Precision (Percentage)
C-1	73	70	77
C-2	76	71	74
C-3	71	69	73
C-4	74	72	77

Figure 3.1
Final Model Confusion Matrix Heat Map

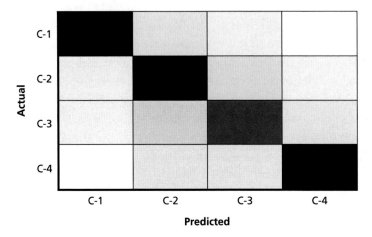

NOTES: Each cell value is normalized by the sum of its row and column (in essence, a matrix of F1 scores), and darker colors indicate a higher number; correct answers appear on the downward diagonal.

We also explored the usefulness of building models for specific types of units. The modeling reported thus far included all types of active component units, organizations, and installations within the service, but we also explored models for specific categories of units to see if they might perform better. One way to group the units is by operational- and institutional-type units, although those are still very heterogeneous groupings. But too small a grouping would result in not having enough training data for the model. We ended up separating out what might be referred to as tip-of-the-spear units in military vernacular; these are what one would think of as the archetypal warfighting units for this service (such as aviation units for the Air Force; infantry, armor, artillery, and aviation for the Army or Marine Corps; and surface, submarine, and aviation units for the Navy). The F1 scores for the models that were estimated using data from only these tip-of-the-spear units were usually very close to the models estimated using all the readiness data for all units in the service. Using only evaluation data for tip-of-the-spear units, the F1 is 0.74 on the model estimated with model data for all units and 0.73 on a comparable model estimated using only model data for tip-of-the-spear units.

Discussion

Overall, the model did well at interpreting natural language descriptions of personnel, equipment, and training factors, as well as other extenuating information, in terms of their readiness implications. Given a few sentences about a unit, the model was able to correctly calculate which of four readiness levels a unit would report 75 percent of the time.

It turned out that the defense-specific word embeddings did not appear to boost model performance. On the contrary, the models that used the pretrained embeddings performed worse. Considerable time and effort went into building those embeddings. Time and effort on building specialized models for tip-of-the-spear units could also be saved, as they provided no boost in predictive performance for these types of units.

In terms of model architecture, the multilayer neural networks as a whole performed much better than the single-layer LR model we used as a baseline. Once a single recurrent layer was added, whether a GRU or an LSTM, model performance further improved, but additional recurrent layers made little or no difference. Smaller batch sizes helped up to a point. The choice of optimizer and learning rate had a small effect on performance as well.

CHAPTER FOUR

Next Steps and Applications

This project has built a model that can interpret unstructured information about a unit in the service and generate a prediction about what its readiness level is. When given a statement such as *Unit is preparing for training exercise next month. We have a large number of incoming noncommissioned officers [or petty officers]. Equipment shortages do not significantly affect mission*, the model will generate probabilities for each C-level. If more information is added—for example, *We are 25 days to T-1*—the probabilities will change. In other words, a person can write a natural language description of the key characteristics, conditions, and activities within the unit and will get an interpretation from the model of how that translates to unit readiness.

A model such as this, or one based on the more recent transformer architecture, could be a component in a bigger system that integrates other models in a way that provides a simulacrum of intelligence—an AI assistant. As a start, it could be integrated into the service's readiness reporting system on an experimental basis to gently prod unit commanders to provide more accurate and detailed comments. This kind of application of AI is used in requirements engineering to improve the clarity of requirements written by nonexpert end users.[1] In the case of

---

[1] Edwin Friesen, Frederik S. Bäumer, and Michaela Geierhos, "CORDULA: Software Requirements Extraction Utilizing Chatbot as Communication Interface," in *Joint Proceedings of REFSQ-2018 Workshops, Doctoral Symposium, Live Studies Track, and Poster Track Co-Located with the 23rd International Conference on Requirements Engineering: Foundation for Software Quality (REFSQ 2018)*, edited by Klaus Schmid, Paola Spoletini,

service readiness, the AI assistant could detect a mismatch between a commander's comments and C-level by comparing the latter with the C-level predicted by the model. It could then ask for greater elaboration to justify the reported C-level. A more advanced system could highlight the few specific words that are most responsible for the mismatch—using word saliency, for example[2]—and prompting the commander to confirm whether these words really convey the intended meaning. Building on word saliency, it might be possible for the system to suggest alternative words that would make the comments better match the reported C-level. The Air Force already conducts such a process of review and correction, albeit manually, via its Major Command Readiness Offices.

Feedback to the commander could be provided in real time, in effect adding a chatbot component to the AI assistant. Various architectures exist to implement chatbots, many techniques using IR or generative models, or a combination of both approaches.[3] Applied to service readiness reporting, an IR technique might work by suggesting alternative versions of the comments that might better fit the reported C-level based on a fixed database (as was used in this work). A generative approach might instead use a model like Seq2Seq to generate conversational sentences.[4] An IR approach would require little

Eya Ben Charrada, Yoram Chisik, Fabiano Dalpiaz, Alessio Ferrari, Peter Forbrig, Xavier Franch, Marite Kirikova, Nazim Madhavji, Cristina Palomares, and Jolita Ralyté, Aachen, Germany: Requirements Engineering: Foundation for Software Quality, 2018.

[2]   Jiwei Li, Xinlei Chen, Eduard Hovy, and Dan Jurafsky, "Visualizing and Understanding Neural Models in NLP," in *Proceedings of the 2016 Conference of the North American Chapter of the Association for Computational Linguistics: Human Language Technologies*, Stroudsburg, Penn.: Association for Computational Linguistics, 2016.

[3]   Minghui Qiu, Feng-Lin Li, Siyu Wang, Xing Gao, Yan Chen, Weipeng Zhao, Haiqing Chen, Jun Huang, and Wei Chu, "AliMe Chat: A Sequence to Sequence and Rerank Based Chatbot Engine," in *Proceedings of the 55th Annual Meeting of the Association for Computational Linguistics*, Stroudsburg, Penn.: Association for Computational Linguistics, 2017.

[4]   Ilya Sutskever, Oriol Vinyals, and Quoc Viet Le, *Sequence to Sequence Learning with Neural Networks*, arXiv, last revised December 14, 2014.

to no additional training data, but would probably yield less-tailored suggestions, whereas the generative approach would provide better-tailored comments but would require collecting written exchanges between a commander and a human assistant to train a model. After initial development and deployment, the accuracy of the system could be further improved during operations by prompting the commander for a simple *good* versus *bad* assessment of the usefulness of the chatbot recommendations and using reinforcement learning to update the model.[5]

Whether via chatbot or a simpler interface, the system developed here could be used both during actual readiness reporting and to train commanders to provide accurate text descriptions. In either case, the model developed here would contribute by assessing the match between the commander's comments and C-level. A similar model trained to predict T-levels instead of C-levels could be used as another means of helping the service develop more objective measures of training readiness. In a sense, a model such as this builds on the wisdom of the crowd. If most commanders tend to say their unit is a C-2 or a T-2 under some particular set of conditions, an individual commander would probably be very interested in knowing that. Such a tool need not even be incorporated into the readiness reporting system; instead, it might be available through the service's training resource website.

Despite their poor performance in the readiness predictive models, the defense-specific embeddings will likely prove useful in other NLP applications that entail matching sequences of text with one another. The task at hand is text classification, which entails matching sequences of text to some categorical outcome. NLP tasks that entail sequence-to-sequence matching include question answering, text summarization, and IR. Each of these tasks requires a model to determine

---

[5] Iulian V. Serban, Chinnadhurai Sankar, Mathieu Germain, Saizheng Zhang, Zhouhan Lin, Sandeep Subramanian, Taesup Kim, Michael Pieper, Sarath Chandar, Nan Rosemary Ke, Sai Rajeshwar, Alexandre de Brebisson, Jose M. R. Sotelo, Dendi Suhubdy, Vincent Michalski, Alexandre Nguyen, Joelle Pineau, and Yoshua Bengio, *A Deep Reinforcement Learning Chatbot*, arXiv, last revised November 5, 2017.

the similarity between two statements such as *The unit has a significant shortage of Class IX supplies* and *The unit is missing many spare parts.* The defense-specific embeddings are well suited to perform such tasks on defense-specific text. Here, again, is an example of transfer learning, in which trained models can be reused in other applications.

Word Embedding Primer

The simplest way to encode text for machine learning tasks is to give each word a unique index, and then encode the words into a 1D vector with a 1 for the element that corresponds to its index and a 0 for all other elements. For example, if the corpus consists of two simple sentences, *The dog bit the man* and *The man ran*, the words could be encoded as shown in Table A.1. This method is known as one-hot encoding.

An entire sentence (or more) could be encoded as the sum of the vectors of its individual words or as a bitwise OR operation. Using the vector sum encoding, *The dog bit the man* would be encoded as [2,1,1,1,0] whereas with the bitwise OR operation it would be [1,1,1,1,0]. There are two major limitations with this approach: the first is that *The man bit the dog* would also be encoded as either [2,1,1,1,0] or [1,1,1,1,0], despite meaning something very different and more interesting; the second is that there is no principled way of conveying any notion of semantics between words like *dog* and *man*: for example, by any distance metric the vector for *dog* is no closer to *man* than it is to *the*, despite that, at a semantic level, *dog* and *man* both denote living entities. Had the second sentence been processed first (or if the man had in fact bitten the dog and that sentence was processed first), the encoding of *man* would have been [0,1,0,0,0], and *dog* would have been something else. In other words, knowing how *man* is encoded tells you nothing about its relationship to any other word in the vocabulary.

There is somewhat more information captured with a TF-IDF encoding, but it still will not tell who is biting whom, and the only relationships that can be inferred by the embedding vectors is the words' relative frequency and usage, not what they mean.

Table A.1
One-Hot Text Encoding

Word	Index	Embedding
'the'	0	[1,0,0,0,0]
'dog'	1	[0,1,0,0,0]
'bit'	2	[0,0,1,0,0]
'man'	3	[0,0,0,1,0]
'ran'	4	[0,0,0,0,1]

To circumvent the limitations of OH encoding, distributed representations of words were proposed at least as far back as in the 1990s. Distributed representations allow all dimensions of a vector—whether there are 5, as in the example above, or 500—to provide information as to the meaning of a word. *Bit* and *ran* are both past-tense verbs, so maybe their vectors should be somewhat similar to one another. A dog and a man are both mammals capable of biting and running, so maybe their vectors should be somewhat similar to one another. Conceptually, the first element of the vector might indicate something about the part of speech—if this is a noun, a verb, or a definite article. The second and third elements might provide further information in the case of a noun—if it is alive, or whether it can move. The fourth and fifth elements could tell something about the word if it represents a verb—if it is in the present tense, or whether it involves movement. If a human were simply encoding these two sentences, the values could still be binary—yes, it is a verb; no, it is not in the present tense; and yes, it involves movement. But with a machine learning approach on, say, the Google News collection of 100 billion words, these parameter values would have to be learned and would unlikely be exactly 0 or 1. This enables each dimension of the vector to represent more fine-grained information about words, with the caveat that the elements of the vectors are no longer interpretable by humans. No element actually means something like a part of speech or the present tense of a verb.

This approach, what Yoshua Bengio and colleagues refer to as distributed representations of words,[1] saw several advances between 2003 and 2013, culminating in the word2vec approach—developed by a team of researchers at Google and introduced in 2013—which is a method of capturing the meaning of words based on their co-occurrence with other words. Word2vec learns to map individual words to vectors in a real vector space. The dimension of the vector space is chosen at the outset and is typically between 100 and 300. This vector representation of a word is referred to as its word embedding. Through training on a large corpus of text, words that tend to co-occur with similar context words will tend to be mapped close to each other in the high-dimensional vector space. The words *car, truck* and *vehicle,* for example are often surrounded by similar context words and hence will be mapped to nearby regions of the vector space. This vector space can thus be considered as encoding some level of semantic information about words. Word2vec enabled machines to perform what at the time seemed like remarkable feats of reasoning by virtue of vector math. For example, the word analogy "X is to Italy what Paris is to France" can be computed with the arithmetic formula *Paris – France + Italy,* which yields the word *Rome.* Geometrically, this corresponds to adding the vector *Paris – France* to the point *Italy* in the word embedding space. Word2vec was followed by other word embedding methods, most notably GloVe.[2] The various word embedding methods exceeded previous state-of-the-art neural network language models on several natural language tasks. Next came huge advances in what are known as transformer models, based on enormous neural networks that can take many days to train even on specialized AI accelerators. A simplified way to think of the difference between word2vec models and transformer models is that the former calculate

---

[1]   Yoshua Bengio, Réjean Ducharme, Pascal Vincent, and Christian Jauvin, "A Neural Probabilistic Language Model," *Journal of Machine Learning Research,* Vol. 3, 2003.

[2]   Jeffrey Pennington, Richard Socher, and Christopher D. Manning, "GloVe: Global Vectors for Word Representation," in *Proceedings of the 2014 Conference on Empirical Methods in Natural Language Processing,* Stroudsburg, Penn.: Association for Computational Linguistics, 2014a.

fixed embeddings for a word based on its average context within a corpus of documents, whereas the latter generate *rules* for calculating embeddings for a word that adapt with every context. The transformer models perform so well that their advent in late 2017 and early 2018 has been termed NLP's ImageNet moment, in the sense that they learn reusable, general-purpose rules for language.[3]

---

[3]   See, for example, Sebastian Ruder, "NLP's ImageNet Moment Has Arrived," webpage, July 12, 2018. Others have also used this term, and it is unclear where it originated.

References

Aydın, Ferhat, Zehra Melce Hüsünbeyi, and Arzucan Özgür, "Automatic Query Generation Using Word Embeddings for Retrieving Passages Describing Experimental Methods," *Database*, Vol. 17, 2017, pp. 1–17.

Bengio, Yoshua, Réjean Ducharme, Pascal Vincent, and Christian Jauvin, "A Neural Probabilistic Language Model," *Journal of Machine Learning Research*, Vol. 3, 2003, pp. 1137–1155.

Chollet, François, *Deep Learning with Python*, Shelter Island, N.Y.: Manning Publications, 2018.

Department of the Air Force, *Force Readiness Reporting*, Washington, D.C.: Department of the Air Force, AFI 10-201, March 3, 2016.

Department of the Army, *Army Unit Status Reporting and Force Registration*, Washington, D.C.: Department of the Army, AR 220-1, April 15, 2010.

Department of the Navy, *Marine Corps Readiness Reporting*, Washington, D.C.: Department of the Navy, MCO 3000.13A, July 18, 2017.

Finkelstein, Lev, Evgeniy Gabrilovich, Yossi Matias, Ehud Rivlin, Zach Solan, Gadi Wolfman, and Eytan Ruppin, "Placing Search in Context: The Concept Revisited," *ACM Transactions on Information Systems*, Vol. 20, No. 1, 2002, pp. 116–131.

Friesen, Edwin, Frederik S. Bäumer, and Michaela Geierhos, "CORDULA: Software Requirements Extraction Utilizing Chatbot as Communication Interface," in *Joint Proceedings of REFSQ-2018 Workshops, Doctoral Symposium, Live Studies Track, and Poster Track Co-Located with the 23rd International Conference on Requirements Engineering: Foundation for Software Quality (REFSQ 2018)*, edited by Klaus Schmid, Paola Spoletini, Eya Ben Charrada, Yoram Chisik, Fabiano Dalpiaz, Alessio Ferrari, Peter Forbrig, Xavier Franch, Marite Kirikova, Nazim Madhavji, Cristina Palomares, and Jolita Ralyté, Aachen, Germany: Requirements Engineering: Foundation for Software Quality, 2018, n.p. As of June 30, 2021: http://ceur-ws.org/Vol-2075/NLP4RE_paper3.pdf

"Gensim: Topic Modelling for Humans," webpage, undated. As of June 30, 2021: https://radimrehurek.com/gensim/index.html

Google Code, "Word2vec," webpage, undated. As of June 30, 2021:
https://code.google.com/archive/p/word2vec/

Hamilton, William L., Jure Leskovec, and Dan Jurafsky, "Diachronic Word Embeddings Reveal Statistical Laws of Semantic Change," in *Proceedings of the 54th Annual Meeting of the Association for Computational Linguistics*, Stroudsburg, Penn.: Association for Computational Linguistics, 2016, pp. 1489–1501.

Joint Staff, *Force Readiness Reporting*, Washington, D.C.: Joint Staff, CJCSI 3401.02B, July 17, 2014.

Khodak, Mikhail, Nikunj Saunshi, Yingyu Liang, Tengyu Ma, Brandon Stewart, and Sanjeev Arora, "A la Carte Embedding: Cheap but Effective Induction of Semantic Feature Vectors," in *Proceedings of the 56th Annual Meeting of the Association for Computational Linguistics*, Stroudsburg, Penn.: Association for Computational Linguistics, 2018, pp. 12–22.

Li, Jiwei, Xinlei Chen, Eduard Hovy, and Dan Jurafsky, "Visualizing and Understanding Neural Models in NLP," in *Proceedings of the 2016 Conference of the North American Chapter of the Association for Computational Linguistics: Human Language Technologies*, Stroudsburg, Penn.: Association for Computational Linguistics, 2016, pp. 681–691.

Mikolov, Tomas, Kai Chen, Greg Corrado, and Jeffrey Dean, *Efficient Estimation of Word Representations in Vector Space*, arXiv, last revised September 7, 2013. As of June 30, 2021:
https://arxiv.org/pdf/1301.3781

Onal, Kezban Dilek, Ye Zhang, Ismail Sengor Altingovde, Md Mustafizur Rahman, Pinar Karagoz, Alex Braylan, Brandon Dang, Heng-Lu Chang, Henna Kim, Quinten McNamara, Aaron Angert, Edward Banner, Vivek Khetan, Tyler McDonnell, An Thanh Nguyen, Dan Xu, Byron C. Wallace, Maarten de Rijke, and Matthew Lease, "Neural Information Retrieval: At the End of the Early Years," *Information Retrieval Journal*, Vol. 21, 2018, pp. 111–182.

Pennington, Jeffrey, Richard Socher, and Christopher D. Manning, "GloVe: Global Vectors for Word Representation," in *Proceedings of the 2014 Conference on Empirical Methods in Natural Language Processing*, Stroudsburg, Penn.: Association for Computational Linguistics, 2014a, pp. 1532–1543.

———, "GloVe: Global Vectors for Word Representation," webpage, Stanford Natural Language Processing Group, August 2014b. As of June 30, 2021:
https://nlp.stanford.edu/projects/glove/

Pruksachatkun, Yada, Jason Phang, Haokun Liu, Phu Mon Htut, Xiaoyi Zhang, Richard Yuanzhe Pang, Clara Vania, Katharina Kann, and Samuel R. Bowman, "Intermediate-Task Transfer Learning with Pretrained Models for Natural Language Understanding: When and Why Does It Work?" in *Proceedings of the 58th Annual Meeting of the Association for Computational Linguistics*, Stroudsburg, Penn.: Association for Computational Linguistics, 2020, pp. 5231–5247.

Qiu, Minghui, Feng-Lin Li, Siyu Wang, Xing Gao, Yan Chen, Weipeng Zhao, Haiqing Chen, Jun Huang, and Wei Chu, "AliMe Chat: A Sequence to Sequence and Rerank Based Chatbot Engine," in *Proceedings of the 55th Annual Meeting of the Association for Computational Linguistics*, Stroudsburg, Penn.: Association for Computational Linguistics, 2017, pp. 489–503.

Ruder, Sebastian, "NLP's ImageNet Moment Has Arrived," webpage, July 12, 2018. As of June 30, 2021:
https://ruder.io/nlp-imagenet/

Serban, Iulian V., Chinnadhurai Sankar, Mathieu Germain, Saizheng Zhang, Zhouhan Lin, Sandeep Subramanian, Taesup Kim, Michael Pieper, Sarath Chandar, Nan Rosemary Ke, Sai Rajeshwar, Alexandre de Brebisson, Jose M. R. Sotelo, Dendi Suhubdy, Vincent Michalski, Alexandre Nguyen, Joelle Pineau, and Yoshua Bengio, *A Deep Reinforcement Learning Chatbot*, arXiv, last revised November 5, 2017. As of June 30, 2021:
https://arxiv.org/pdf/1709.02349

Sutskever, Ilya, Oriol Vinyals, and Quoc Viet Le, *Sequence to Sequence Learning with Neural Networks*, arXiv, last revised December 14, 2014. As of June 30, 2021:
https://arxiv.org/pdf/1409.3215